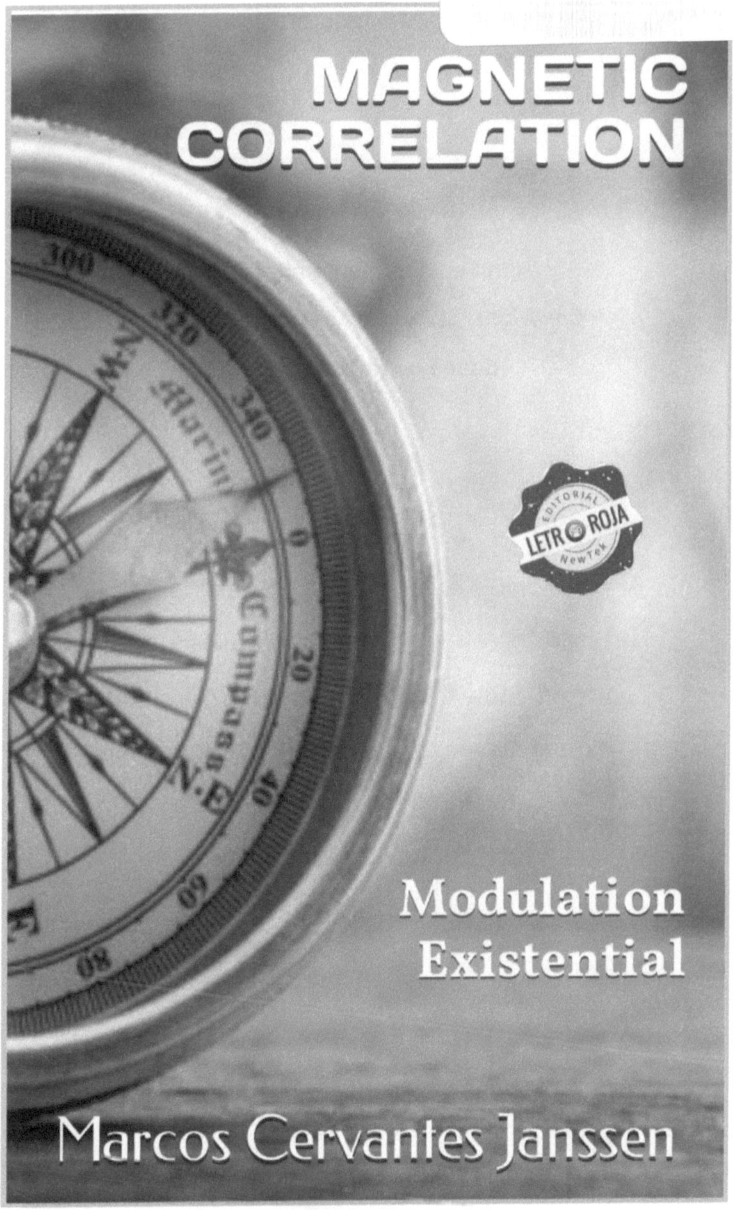
MAGNETIC
CORRELATION
EDITORIAL
LETRA ROJA
NewTek
Modulation
Existential
Marcos Cervantes Janssen

First edition: September 7, 2023

Copyright © *2023 Marcos Cervantes Janssen*

Edited by Editorial letr@red

https://www.youtube.com/channel/UCQ12Xlt8oQOaWAhAiboXPUA

https://www.instagram.com/newtekjanssen/

https://www.facebook.com/LETRA3ROJA

https://www.newtek.janssen@gmail.com

https://twitter.com/Letra3Roja

https://newtekjanssen.es.tl/

letra3roja@gmail.com

LETR ROJA.

MAGNETIC CORRELATION

Modulation
Existential

By: Marcos Cervantes Janssen.

INDEX:

FOREWORD:

The existential space-energy relationship only has form and circumstance due to the magnetic force that relates it; This is what this writing focuses on, plus the collateral contributions will be of great interest for the complete study of the topic.

Therefore, I first exist, and for this reason I am conscious when thinking.

Existence is what includes relative individuality, in absolute eternity, an absurdity for temporality, more a reality than theological for eternity.

Polarization in magnetism gives life to the expression, "Magnetic Modulation." This being the form that defines the energy contained in it; Let's say, the mind of a physical structure is called magnetic modulation.

The magnetic correlation of everything that inhabits this existence is visible in the constant kinetics of movement, and this modulation is called evolution, when its systems take more efficient forms through existential chronology.

Obeying a progressive evolution is what determines the magnetic correlation as expansive and inclusive in everything that is manifested.

In this writing we will discuss the magnetic nature of existence, and its modulation as a dynamic evolutionary conscious, so I invite you to pay attention to the abstract message anddiscern with complete freedom. We will look for the magnetic correlation that exists to modulate our existence, at studyable and at the same time non-studyable levels, within the scope of creative intuition.

1 - THE CORRELATION:

Correlation is a statistical measure that indicates the relationship between two variables. It is used to determine whether a relationship exists between two sets of data and, if so, what type of relationship it is (positive, negative, or null).

The correlation can be calculated using different methods, such as Pearson's correlation coefficient or Spearman's correlation coefficient.

In general, the closer the correlation value is to 1 or -1, the greater the relationship between the variables, while a value close to 0 indicates that there is no relationship between them.

Magnetic correlation refers to the relationship between the magnetic signal measured on a magnetic resonance image (MRI) and the anatomical structure of the tissue.

The magnetic signal is produced from the interaction between magnetic fields and protons in the tissue.

Magnetic correlation is used in the interpretation of MRI images to identify different types of tissue and anatomical structures.

For example, magnetic correlation can help identify tumors or lesions in the brain or other organs.

The correlation **neutral** It refers to the absence of a relationship between two variables. In other words, when the correlation between two sets of data is close to zero, it can be said that there is no significant relationship between them.

This can be useful in some cases, as it may indicate that certain variables are unrelated and therefore do not need to be considered together in an analysis or model.

The correlation**negative** refers to an inverse relationship between two variables, meaning that when one variable increases, the other variable tends to decrease.

An example of a negative correlation could be the relationship between sleep time and stress level.

If there is a strong negative correlation between these two variables, then it is likely that people who sleep fewer hoursexperiments higher stress levels.

The correlation**positive** refers to a direct relationship between two variables.

This means that when one variable increases, the other also tends to increase, and when one variable decreases, the other also tends to decrease.

In other words, both variables move in the same direction.

An example of a positive correlation could be the relationship between the number of hours of study and the grades obtained on an exam: as the number of hours of study increases, the grades obtained also increase.

It is in this way that, knowing what correlation means, we understand in existence the importance of relating with those who seem to be totally contrary to us.

It is interesting how in all of our reality, mathematics helps us understand not only the material world, but also the emotional and mental world in which we live.

2 - MAGNETISM:

Magnetism is a fundamental force found throughout the universe and is essential to understanding many cosmic phenomena.

Magnetism is present in stars, planets, galaxies and other celestial objects.

For example, Earth's magnetic field is what protects us from solar and cosmic radiation, while in stars, magnetism can generate solar flares and other violent events.

In addition, magnetic fields can also influence the formation and evolution of cosmic structures, such as galaxies and galaxy clusters.

In summary, magnetism is a fundamental force that plays an important role in the universe and its study is essential to understand many cosmic phenomena, as well as life itself on this beautiful planet.

Now, in a thinking and intelligent universe, gravity also has a personal behavior, thus through psychology we can understand our existence in a comprehensive way, and become personally intimate with the existence in which we inhabit, in an infinite set.

The term "psychological magnetism" refers to a person's ability to influence the emotions, thoughts, and behaviors of others through their presence, body language, communication skills, and other psychological techniques.

Psychological magnetism can be used to establish healthy and effective interpersonal relationships, as well as to persuade others to adopt a certain opinion or behavior.

However, it can also be used in a manipulative or abusive manner, so it is important to use this skill responsibly and ethically.

Thus, magnetism is a phenomenon not only spatial or physical, but also psychological, emotional and managed in all areas of study, whether scientific or even esoteric.

Quantum physics reveals a strong correlation between scientific magnetism and the electro-spatial vibration of our neurons when thinking, this study is exciting and powerful.

3 - NEURAL AND SPACE TISSUE:

Our neurons are arranged as a highly communicated tissue, that is, with direct correlation, constant and flexible in nature. A strong flow of energy gathers, through forces finally known today, as an electromagnetic mental field.

This structural field of energetic data takes place physically in the coming and going of our neurotransmitters, generating an energetic mass and a mental reality in which we inhabit, to develop as true humans.

I emphasize the spatial fabric, with its enormous similarity to our mind, for sharing the same root structure, which is expansive, and which seems to have no limit.

In the same way that the human mind evolves in expansion, the universes expand to infinity and we will call this wonderful procedure in this essay as existential modulation.

Well, the defined and extraordinary formula that is executed for this purpose will be of incredible and complex dimensions.

The visible part of this matterit seemed clear and of a perfect order, plus the diversity of infinite forms will always be chaos for human reason due to its complexity, even though it is of perfectly ordered eternity.

We will take the material and mental part of existence as an evolving organic fabric.

Neural and spatial tissue refers to the organization and distribution of nerve cells in the brain and its relationship to cognitive and spatial functions.

Neural tissue is made up of different types of nerve cells, including neurons and glial cells, that work together to process information and carry out cognitive functions such as memory, learning and perception.

Spatial fabric, on the other hand, refers to the way the brain processes and represents spatial information, such as the location of objects in the environment and navigation.

Neural and spatial tissue are closely related and work together to enable the processing of complex information and the performance of complex cognitive tasks.

4 - MAGNETIC TIME:

The times in which magnetism acts determine the speed of evolution, the concept magnetic time It is not handled, however in this writing I will give it a personal interpretation, for the understanding and study of the magnetic relationship with modulation.

Magnetism marks structural lines that fluctuate in the spatial conformation, but over time we must observe their movements and new formations.

The static only exists in times of very long periods relative to others.

Magnetic time defines the modulation obtained in an expansional line of light, plus the slopes and diversities of its forms play an eternal role, called relative destiny.

Magnetism over time refers to the variation of the magnetic field over time.

The Earth's magnetic field, for example, has undergone significant changes throughout geological history, and these changes can be detected and studied through geological and paleomagnetic records.

Additionally, magnetism can also be used to date rocks and other geological materials through the technique known as paleomagnetism dating.

In summary, magnetism over time is an important concept in geology and physics, and its study can provide valuable information about the geological history and evolution of our planet.

4 - MAGNETIC MODULATION:

Magnetic modulation is the form that matter takes, through magnetic lines predisposed by an existential intelligence that makes up everything, every movement of energy in the universe obeys this modulation, including the creative thoughts of all beings involved in this wonderful action.

The word "modulation" comes from the concept of "shaping", and similarly, electrical energy is shaped into an infinite number of electro-spatial flows known as magnetic arrays.

Through this magnetic modulation process, information is transmitted and manipulated efficiently by varying the amplitude of the magnetic signal.

The essence of existence is perpetual creation, based on eternal transformation, known as evolution. For science, magnetic modulation is a signal coding technique used in data transmission.

It consists of varying the amplitude of a high-frequency magnetic signal to represent digital information.

Magnetic modulation is used in various applications, such as magnetic tape recording and wireless data communication in industrial control and automation systems.

It is interesting to think about how energy and magnetic lines can be seen as a way of shaping matter and how everything in the universe is connected through this modulation.

It is also true that evolution and transformation are fundamental concepts in existence, which is why they are worthy of our study.

Let us understand neuronal functioning as an electronic transfer in space, thus our minds being biological generators of very high precision and constant activity.

The responsibility is ours, because today we know that our thoughts affect our surroundings, the distance, frequency and power, differs due to multiple internal or external factors of each living being in this large group of thinking beings.

Let's take care of the input peripherals, ears, touch, taste, smell and vision, as well as the output, mouth, extremities and above all celebrate with your thoughts.

6 - EXISTENTIAL CORRELATION:

Everything and everyone in this existence have energy lines in common that unite us infinitely, it is then where the form responds to a single expanding mind.

Our mission as thinking beings is to synchronize among ourselves tothen wake up continually to the unified reason of the whole, it is here that individual freedom ends with the existential subjection of the evolutionary flow.

The only path in which everything begins and ends cyclically is the very nature of existing in an infinitely diverse form of life in the eternity of ordered chaos, which always existed creating times as an eternal evolving chronology.

Existential correlation is a term that refers to the interconnection and mutual dependence between all forms of life and nature on the planet.

This idea suggests that all forms of life are interconnected and that every action we take affects everything else in the natural world.

Existential correlation is important because it reminds us that we are part of a larger ecosystem and that our actions have consequences on the world around us.

It is important to take this interdependence into account when making decisions and acting in a responsible and sustainable manner for the well-being of the planet and all forms of life that inhabit it.

EPILOGUE:

Correlation is a statistical measure that indicates the relationship between two variables and is used to determine whether a relationship exists between two sets of data and what type of relationship it is.

Magnetic correlation refers to the relationship between the magnetic signal measured in an MRI image and the anatomical structure of the tissue, which helps identify different types of tissue and anatomical structures.

Magnetic modulation is a signal coding technique used in data transmission and consists of varying the amplitude of a high-frequency magnetic signal to represent digital information.

Neutral correlation refers to the absence of relationship between two variables, while negative correlation refers to an inverse relationship between two variables.

These concepts are interconnected and are applied in different areas of study, such as physics, psychology and medicine.

After this, as important information, I will tell you that magnetic correlation is of vital importance for an existential modulation, because without any relationship, the particles in existence are isolated and remain at rest until they are part of an evolving life system.

I will say without a doubt that there are only two types of energy, kinetic and aesthetic, the first being existence and the second being the static origin of it.

AS A TELECOMMUNICATIONS ENGINEER, THE CORRELATION BETWEEN PARTICLES, DENOTES IN MY LIFE A CONSTANT COMMUNICATION OF EXISTENCE, IN MY PERSONAL EXPERIENCE, I ASSURE YOU THAT YOUR THOUGHTS INFLUENCE AND ARE INFLUENCED BY THE TOTAL AROUND YOU, I INVITE YOU TO ENTER INTO EXISTENTIAL COMMUNION .